S0-AJZ-826

1. 19 - 5 = _____

2. 20 + _____ = 27

3. Extend the following pattern.

  3, 6, 9, 12, _____, _____, _____

4. Is this a growing, shrinking or repeating pattern?

  30, 27, 24, 21, 18

5. Create a pattern based on the following rule: start at 10, add 5

TUESDAY — Number Sense

1. Elizabeth is 7ᵗʰ in line to get a drink after gym. How many people are in front of her?

  _____

2. Order these numbers from least to greatest.

  8791, 7918, 7911

  _____, _____, _____

3. What is the greatest number you can make from these digits?

  3   6   1   5   _____

4. Round these numbers to the nearest ten.

  456 _____   284 _____

5. Multiply: 8 X 5 =

WEDNESDAY — Geometry and Spatial Sense

1. Name this shape.

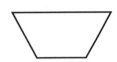

_____

2. How many sides does a polygon have?

3. Name this 3D figure.

_____

4. Look at the shapes. Choose flip, slide or turn.

 →

    A. flip   B. slide   C. turn

5. What is a quadrilateral?

_____

THURSDAY — Measurement

1. How many mm in one cm?

_____

2. What time is it?

 _____

3. Should you measure a pencil with m, cm or km?

_____

4. What does km stand for?

_____

5. Perimeter = _____ units

   Area = _____ square units

Data Management

The grade four classes took a survey of their favourite sports. They displayed the data as a pictograph. Use their pictograph to answer the questions.

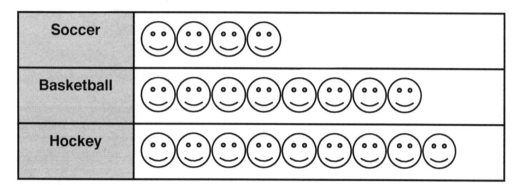

Soccer	
Basketball	
Hockey	

 = 2 students

1. How many students took part in the survey? _____

2. What was the most popular sport? _____

3. How many more students preferred hockey over basketball?

_____

4. List the favourite sports from least to greatest.

_____

5. What would be a good title for this pictograph?

_____

BRAIN STRETCH

1. How many weeks are there in one year? _____

2. How many weeks are there in two years? _____

3. How many weeks are there in five years? _____

MONDAY — Patterning and Algebra

1. ___ + 5 = 20

2. Complete the following pattern:

   5, 10, 15, 20, _____, _____, _____

3. Complete the following pattern:

   11, 22, 11, 22, _____, _____, _____

4. Complete the following pattern:

   80 ,70 ,60 ,50 , _____, _____, _____

5. Divide:

   72 ÷ 8 =

TUESDAY — Number Sense

1. Write the numeral.

   twelve thousand two hundred five

2. Order these numbers from least to greatest.

   4723, 7432, 7324

   _____, _____, _____

3. What is the greatest number you can make from these digits?

   1 6 4 5 _____

4. Round these numbers to the nearest hundred:

   1263 _____   2730 _____

5. What is the value of the money?

1. Name this shape.

_____

2. How many parallel sides does a square have?

3. Name this 3D figure.

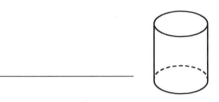

_____

4. Look at the shapes. Choose flip, slide or turn.

   A. flip   B. slide   C. turn

5. Name a quadrilateral.

_____

THURSDAY — Measurement

1. How many m in a km?

2. What temperature is it most likely if you are in a bathing suit at the beach?

   A. 30°C   B. - 3°C   C. 6°C

3. How long would it take to write the date on a piece of paper?

  A. 1 min   B. 1 day   C. 1 hour

4. What time is it?

 _____

5. Would you measure the height of a house with m, cm or mm?

Sarah's music class took a survey of students in her school to see what kind of music they preferred to listen to. Here are her results: 12 preferred reggae, 2 classical, 28 dance music, 17 rock'n roll, and 14 preferred country.

Reggae	Classical	Dance	Rock'n Roll	Country

1. Use the table above to show a tally of the results.

2. What was the most popular style of music? _____

3. What was the least popular style of music? _____

4. How many more students preferred dance music to country music?

   _____

5. How many students participated in this survey? _____

BRAIN STRETCH

John has 1268 marbles and wants to give his sister half of them.
How many will he give her?

MONDAY — Patterning and Algebra

1. 50 − 25 =

2. What kind of pattern is this?

   12, 24, 36, 48

   A. growing   B. repeating   C. shrinking

3. 12 + 10 =

4. 20 - _____ = 12

5. Continue the pattern:

   11, 22, 33, 44, _____, _____, _____

TUESDAY — Number Sense

1. What number is 1000 more than 7983?

2. How many thousands in 5746?

3. Multiply:

   65 X 10 =

4. Write the numeral for:

   eight hundred sixteen

5. What is the value of the money?

WEDNESDAY — Geometry and Spatial Sense

1. Name this shape.

   _____

2. How many parallel sides does a diamond have?

   _____

3. Name this 3D figure.

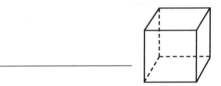

   _____

4. What shape is not a quadrilateral?

   A. octagon

   B. rhombus

   C. square

5. How many sides and vertices does an octagon have?

   sides _____   vertices _____

THURSDAY — Measurement

1. 1 m = _____ cm

2. 2 dm = ___ cm

3. What unit of measurement would you use to measure the distance from your house to school?

4. What time is it?

5. What is the area covered?

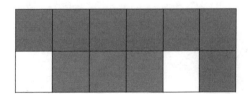

   _____ square units

Ms. Jenkins wanted to make a community quilt and asked her neighbours what their favourite colours were so that she could include them in her design. She made a graph from the data she collected.

Title_____

Favourite Colour

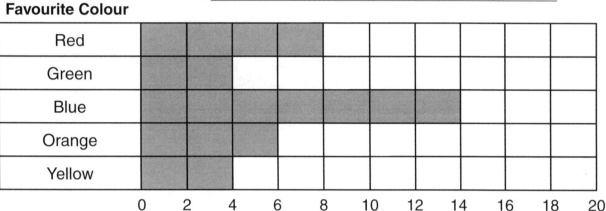

Red										
Green										
Blue										
Orange										
Yellow										

0   2   4   6   8   10   12   14   16   18   20

1. Label the graph.

2. What is the most favourite colour? _____

3. Which two colours are least favourite? _____

4. How many people were asked in this survey?_____

BRAIN STRETCH

If you had to build a tower with only one type of 3D figure, what 3D figure would you use? Why?

MONDAY — Patterning and Algebra

1. What kind of pattern is it?

   2, 4, 6, 8, 10, 12, 14

2. What kind of pattern is it?

   2, 2 ,2, 3, 3, 3, 4, 4, 4, 5, 5, 5

3. What kind of pattern is it?

   24, 22, 20, 18, 16

4. $45 - 10 =$

5. Complete the following:

   $12 + \_\_\_ = 23$

TUESDAY — Number Sense

1. Order these numbers from least to greatest.

   2745, 2543, 3245

   _____, _____, _____

2. Order these numbers from least to greatest.

   5208, 8052, 5028

   _____, _____, _____

3. Subtract:

   $\begin{array}{r} 6605 \\ - 3252 \\ \hline \end{array}$

4. Multiply:

   $\begin{array}{r} 45 \\ \times 4 \\ \hline \end{array}$

5. What is the greatest number you can make from these digits:

   8 6 1 2

WEDNESDAY — Geometry and Spatial Sense

1. Name this shape.

_____

2. What is the name of a polygon with eight sides?

_____

3. How many lines of symmetry?

K _____

4. How many vertices does a triangle have?

_____

5. What is the name of this 3D figure?

 _____

THURSDAY — Measurement

1. 6 cm = _____ mm

2. 7000 mm = _____ m

3. What unit of measure would you use to measure the width of your finger?

4. It is 7:50 am. What time will it be in 40 minutes?

5. Find the perimeter and area of the rectangle.

6 cm

2 cm

P = _____

A = _____

Data Management

The graph below shows the average yearly precipitation in seven different towns

Average Yearly Precipitation

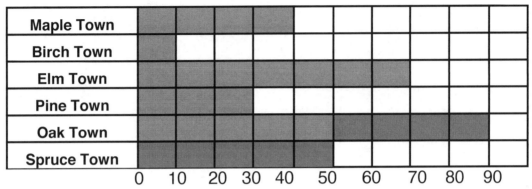

Name of Town

Amount Of Precipitation In Milimetres

1. Which city had the most precipitation? _____

2. Which city had the least precipitation? _____

3. What is the range of the data shown? _____

4. How much more precipitation did Oak Town have than Pine Town?

_____

5. List the towns in order from least precipitation to the most precipitation.

_____

BRAIN STRETCH

What is the perimeter of a rectangular field that is 3 m by 4 m? What is the perimeter in cm?

1. Create a pattern based on the following rule:

   start at 2, add 2

2. Create a pattern based on the following rule:

   start at 1, add 5

3. Create a pattern based on the following rule:

   start at 20, subtract 2

4. Create a pattern based on the following rule:

   start at 3, add 3

5. What is the number that completes the following equation?

   7 + 9 = 8 + _____

TUESDAY — Number Sense

1. Add:

   5778
   + 3641
   ‾‾‾‾‾‾

2. Round to the nearest ten.

   5281

3. What number is 1000 more than 4821?

   _____

4. Choose: >, <, or =.

   128 ☐ 128

5. Make the greatest possible number with these digits:

   7 3 1 4

1. Name this shape.

_____

2. What is the name of a polygon with five sides?

_____

3. How many lines of symmetry?

 _____

4. How many vertices does a pentagon have?

_____

5. What is the name of this 3D figure?

 _____

1. How many decades in 3 centuries?

2. Find the perimeter of a square that has 12 m sides.

3. 900 mm = _____ dm

4. What time is it?

5. Mary wants to build a fence around her rectangular garden. It is 10 m wide and 15 m long. How much fencing does she need to buy?

Data Management

Sam and Shawn surveyed the traffic that passed by their apartment building during the lunch hour. They collected their data using the tally below.

Type of Vehicle	Tally	Number
Cars	~~||||~~ ~~||||~~	
Trucks	||||	
Vans	~~||||~~ ~~||||~~ |||	
Buses	~~||||~~ ~~||||~~ ~~||||~~ ~~||||~~ ||	

1. How many vehicles passed their building? _____

2. Which was the most common vehicle? _____

3. How many more vans than cars passed them? _____

4. How many fewer trucks than buses passed them? _____

5. How many more buses than trucks passed them? _____

BRAIN STRETCH

Mr. Turnbull's class raised $3100 to buy new computers. If a computer costs $800, how many computers can they buy?

MONDAY — Patterning and Algebra

1. Create a pattern based on the following rule:

   start at 30, add 5

2. Create a pattern based on the following rule:

   start at 100, subtract 20

3. What is the missing number?

   $81 \div \boxed{} = 9$

4. What is the pattern rule?

   8, 16, 32, 64

5. What is the number that completes the following equation?

   14 - 8 = 13 - _____

TUESDAY — Number Sense

1. Estimate the sum:

   244 + 122

2. Estimate the sum:

   899 + 501

3. Add:

   124 + 355

4. Add:

   456 + 331

5. 147 students attended the fair in the day and 234 students attended at night. How many students attended in total?

1. What is a vertex?

2. How many sides does a parallelogram have?

3. Draw the lines of symmetry for this square.

4. What 3D figure does this object look like?

5. Draw the reflection of this shape:

THURSDAY — Measurement

1. Draw a rectangle with an area of 24 units^2.

2. What is the area of this shape?

7 units

4 units

3. Draw rectangle with a perimeter of 12 units.

4. What time is it?

5. How many minutes in 4 hours?

Data Management

During the Annual Community Pet Show, Martha surveyed the contestants. Use the space below to create a bar graph of Martha's data. Here are her results:

30 cats   22 dogs   5 ferrets   10 parrots   8 guinea pigs

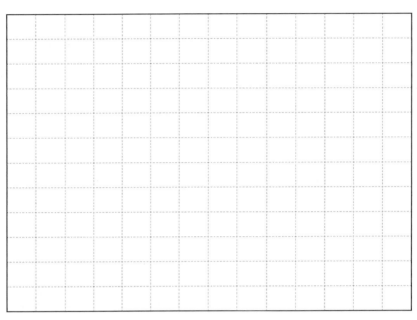

1. What was the most popular pet? _____

2. How many pets came to the show? _____

3. How many more dogs than guinea pigs were there? _____

4. What is the range of her data? _____

BRAIN STRETCH

Marlene's mom lets her play video games for 30 minutes every night. How many hours does she play in a week? How many hours does she play in a month?

MONDAY — Patterning and Algebra

1. Complete the following equation:

   12 + 10 = 15 + ____

2. Complete the following equation:

   5 + 2 = 6 + ____

3. Complete the following equation:

   25 − 6 = ____ - 11

4. Complete the following equation:

   89 = 21 + ____

5. Create a repeating pattern.

TUESDAY — Number Sense

1. Estimate the difference:

   100 - 21

2. Estimate the difference:

   405 - 220

3. Subtract:

   346 - 232

4. Subtract:

   568 - 447

5. Sheila had $90. She bought two books for $18 each. How much money does she have left over?

WEDNESDAY — Geometry and Spatial Sense

1. Draw a parallelogram.

2. How many more sides does an octagon have than a triangle?

3. Can a cube roll?

4. Circle the quadrilaterals.

4. What shape is the face of a cone?

THURSDAY — Measurement

1. Name an object shorter than a decimetre.

2. 8 m = _____ mm

3. What unit of measure would you use to measure a drop of rain?

4. It is 2:35 pm. What time will it be in 45 minutes?

5. Find the perimeter and area of the rectangle.

9 cm

7 cm

P = _____

A = _____

Sam had 10 marbles in a bag: 3 brown and 7 yellow.

1. How likely is it that he will choose a brown marble?

2. How likely is it that Sam will choose a yellow marble?

3. What is the probability of choosing a brown marble?

4. What is the probability of choosing a yellow marble?

5. How likely is it that you will watch television today?

BRAIN STRETCH

The Northland Science Reserve has many fields in which their animals can feed and play. Small farm animals have a field measuring 15 m by 10 m.

1. What is the perimeter of the field?

2. What is the area of the field?

3. If it costs $10 for each metre of fencing, how much will it cost to buy a new fence for the field?

MONDAY Patterning and Algebra

1. Use the rule to make a pattern:

   start at 4, multiply by 4

2. Use the rule to make a pattern:

   start at 2, multiply by 3

3. What is the missing number?

   $\boxed{} \div 7 = 7$

4. What is the pattern rule?

   99, 88, 77, 66

5. Create a growing pattern.

TUESDAY Number Sense

1. Add:

   5672 + 367

2. Add:

   599 + 405 =

3. Subtract:

   74 - 34

4. Subtract:

   68 - 49

5. Mrs. Stevens' class used 188 notebooks from a box of 250.
   How many were left?

WEDNESDAY — Geometry and Spatial Sense

1. A square does not have right angles.

   true   or   false

2. Which of the following 3D figures does not roll?

   A. cylinder B. cone C. cube

3. How many edges does a pyramid have?

4. Draw a net for a cube.

5. Circle the following pairs of lines that intersect.

THURSDAY — Measurement

1. What is the perimeter of a pentagon with sides each 4 cm long?

2. 2 km = _____ cm

3. What is longer?

   A. 100 cm   B. 10 m   C. 1 km

4. How many years in 3 centuries?

5. What is the area covered?

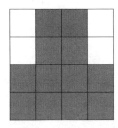

1. When rolling a die, what is the probability that you will roll a 2?

2. What is the probability of rolling a 3?

3. What is the probability of rolling an even number? An odd number?

4. What is the probability of rolling a 7?

5. What is the value of the tally marks?

BRAIN STRETCH

A jar can hold 500 ml of jam.

1. Express that amount in litres.

2. How many jars will you need to have 6.5 L of jam?

3. How much will the jam in question 2 cost if each jar is $2.00?

MONDAY Patterning and Algebra

1. Complete the pattern:

   1, 3, 9, 27, _____, _____, _____

2. Complete the pattern:

   4, 8,12,16, _____, _____, _____

3. Complete the pattern:

   6,12,18, _____, _____, _____

4. Complete the pattern:

   20, 18, 16, 14, _____, _____, _____

5. Create a shrinking pattern.

TUESDAY Number Sense

1. Name the factors for 24.

2. Divide:

   $9\overline{)21}$

3. Write 45 023 in expanded form.

4. What number is 1000 less than 3512?

5. Write the following in standard form:

   2000 + 400 + 60 + 5

1. Classify the following pair of lines.

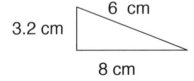

    A. intersecting   B. parallel   C. perpendicular

2. Classify this triangle.

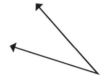

3. An angle of 150° is called?

    A. obtuse   B. right   C. acute

4. Classify this angle.

5. Name a 3D figure that does not have edges.

1. 200 mm = ____dm

2. 16 km =_____m

3. Find the perimeter of the figure:

    6 cm

3.2 cm

    8 cm

4. How many decades in 3 centuries?

5. Find the area of a square that has 3 m sides.

Data Management

Mr. Ramesh created a graph showing how he and his family spent their holiday.

1. What kind of graph did he create?

2. Where did they spend most of their vacation?

3. What sorts of activities did the Ramesh family do on their vacations?

4. What did they do the least amount of time?

5. Where do you think they might have gone on their holiday?

BRAIN STRETCH

Rick is older than John. John is older than Betty. Betty is older than Tina, and Vivienne is older than Rick. Who the oldest?

MONDAY Patterning and Algebra

1. Are the sums in each pair equal?

   $2 + 6$ $1 + 7$

2. Complete.

   $16 - 7 = 18 -$ _____

3. Write 10 as the sum of 3 numbers.

4. Predict what the 21st figure will be in this pattern.

   A. 🚲   B. 🚑   C. 🚒

5. If each bag holds 15 jelly beans. How many jelly beans will there be in 6 bags?

TUESDAY Number Sense

1. Which digit is in the hundreds place?

   4955

2. Use the digits 8,9,3,2 to create the greatest number possible.

3. Subtract:

     7830
   − 6782

4. Order these numbers from greatest to least.

   1751, 7450, 4531, 4879

   _____, _____, _____, _____

5. What is the value of the money?

1. How many lines of symmetry?

2. What is the name of this shape?

3. How many vertices does this shape have?

4. What 3D figure could be made from these pieces?

A. cylinder
B. rectangular prism
C. pyramid

5. Name a type of triangle.

THURSDAY — Measurement

1. What time is it?

2. Kate and Mark went for a 6 km hike. How many metres did they hike?

3. Find the perimeter of the octagon.

5

4. What is the year 1 decade after 1974?

5. 3000 m = _____ km

Data Management

Darcy celebrated her ninth birthday at a local fairground. She invited her seven closest friends. She took a survey of their heights to see which rides they could go on. Here are her results:

Samantha 124 cm

Martin 146 cm

Darcy 130 cm

Larissa 129 cm

Josie 109 cm

Martha 140 cm

Jon 122 cm

Michael 145 cm

1. There is a 120 cm height requirement for the go-carts. Who can ride them?

2. Only those 130 cm or taller can go on the roller coaster. Who is going for a ride?

3. Children 100 cm or taller can go on the ferris wheel. Who can go on that ride?

4. What is the range of the data?

BRAIN STRETCH

Lillian ran 1.4 km on Monday, 1.8 km on Wednesday and 2.1 km on Friday. How many km did she run altogether? How much farther did she run at the end of the week than at the beginning?

MONDAY — Patterning and Algebra

1. Complete the pattern:

   4, 14, 24, _____, _____, _____

2. Complete the pattern:

   20, 40, 80 , _____, _____, _____

3. Each box holds 12 cookies. How many cookies in 6 boxes?

4. Find the missing number.

   200, 175, _____, 125, 100

5. Create a repeating pattern.

TUESDAY — Number Sense

1. Put these fractions in order from greatest to least:

   $\frac{2}{8}$ $\frac{4}{8}$ $\frac{7}{8}$

2. Put these fractions in order from greatest to least:

   $\frac{2}{5}$ $\frac{1}{5}$ $\frac{3}{5}$

3. What fraction of this pie is shaded?

4. If Sue gave Jim half of an 8 piece chocolate bar, how many pieces would she have to give him?

5. Draw $3.20 using 5 coins.

WEDNESDAY   Geometry and Spatial Sense

1. How many lines of symmetry?

2. Name a polygon that has six vertices.

3. Name a polygon with 2 pairs of parallel lines.

4. Which pair of figures is congruent?

A.   B.

5. How many more vertices does a quadrilateral have than a triangle?

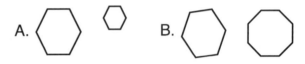

THURSDAY   Measurement

1. What time is it?

2. The time is 2:15 pm. What time will it be in 55 minutes?

3. What is the perimeter of a rectangle 5 cm wide and 8 cm long?

4. What is the year 1 decade after 1934?

5. About how long would it take to brush your teeth?

A. 4 mins   B. 4 seconds   C. 4 hours

Data Management

Everyone at Michael's school participated in a Winter Activity Day. Each person could choose the activity the wanted to do. This bar graph shows their choices.

Activity										
Snowboarding										
Skiing										
Snowshoeing										
Ice Fishing										
Skating										

0　5　10　15　20　25　30　35　40　45　50

Number of Students

1. How many students participated in winter activity day? _____

2. What was the most popular choice? _____

3. What was the least popular choice? _____

4. What is the range of the data? _____

5. How many people chose either skating or skiing? _____

BRAIN STRETCH

Leonard delivers newspapers to 21 houses on Main St., 16 houses on Elm St. and 34 houses on Lincoln Ave. How many papers does he deliver altogether each day? How many papers does he deliver in a week?

MONDAY Patterning and Algebra

1. Create a repeating pattern:

2. $16 = 4 \times$ _____

3. Complete the following:

    11, 22, 33, 44, _____, _____, _____

4. Fill in the missing numbers.

    4, 8, _____, 16, 20, 24, _____,

5. Luke had 27 marbles. He put the marbles in groups of 9. How many equal groups of marbles were there?

TUESDAY Number Sense

1. Put these fractions in order from least to greatest.

    $\dfrac{4}{5}$ $\dfrac{1}{5}$ $\dfrac{3}{5}$

2. Write four tenths in decimal form.

3. Write the following decimal as a fraction.

    0.57

4. What fraction of the pie is not shaded?

5. How much money?

    3 twenty-dollar bills, 3 toonies, 1 loonie, 3 quarters and 5 pennies

WEDNESDAY — Geometry and Spatial Sense

1. Classify the following angle.

2. How many sets of parallel lines does a triangle have?

3. How many vertices does a parallelogram have?

4. What is a polygon?

5. Which of the line segments on the trapezoid are parallel?

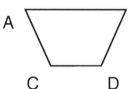

THURSDAY — Measurement

1. What time is it?

2. The time is 4:05 pm. What time will it be in 55 minutes?

3.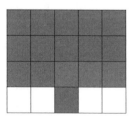

Perimeter = _____ units

Area = _____ square units

4. Each week Ben plays soccer for 530 minutes. How many hours and minutes does he play soccer?

5.   6 km = _____ cm

Describe an event to match each probability:

1. impossible _____

2. unlikely _____

3. very likely _____

4. certain _____

BRAIN STRETCH

Melissa's family is traveling by train to Lewistown to see their aunt.

1. If each car is 11 m long, and there are 9 cars in their train, how long is the train they are traveling on?

2. Each car can hold 48 passengers. How many people can ride on the train at one time?

3. If they left their town at 9 am and are going to travel for 7 hours, when will they arrive at their destination?

MONDAY Patterning and Algebra

1. Create a shrinking pattern:

2. What are the next 5 terms in the pattern?

   1, 2, 2, 3, 3, 3, 4, 4, 4, 4,

3. Fill in the missing number in the pattern:

   19, 17, ___, 13, 11, 9

4. Is this a shrinking pattern?

   65, 60, 55, 50

5. Fill in the missing number in the pattern:

   11, 22, 33, ___, 55

TUESDAY Number Sense

1. Write ninety dollars in decimal form.

2. Write four hundred eighty dollars in decimal form.

3. Draw the money needed to represent $8.50.

4. If you had three ten dollar bills and five loonies, how much money would you have?

5. What is the value of the money?

1. How many lines of symmetry?

L

2. What kind of lines do these line segments show?

3. Draw a pair of congruent shapes.

4. Classify the triangle as equilateral, isosceles or scalene.

5. Look at the shapes. Choose flip, slide, or turn.

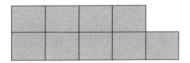

1. Find the area:

A = _____

2. What time is it?

3. Draw a pencil that is three centimetres long.

4. A book's mass is most likely:

    A. 500 g

    B. 500 kg

    C. 5 t

5. 1 kg = _____ g

Data Management

The Eagles soccer team won every game this season. Here is a list of their scores:

1st game - 5	3rd game - 4	5th game - 2	7th game - 3
2nd game - 2	4th game - 2	6th game - 5	8th game - 2

1. What is the range of this data?

2. What is the most common score they got?

3. What is the highest score?

4. What is the lowest score?

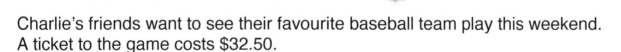

BRAIN STRETCH

Charlie's friends want to see their favourite baseball team play this weekend. A ticket to the game costs $32.50.

1. How much will it cost for four people to attend the game?

2. If the game begins at 1:00pm and it takes the boys 1.5 hours to get to the field, when should they leave their houses?

3. If a program costs $5.25, how much change will someone get if they use a $10 bill to pay for it?

MONDAY   Patterning and Algebra

1. Create a growing pattern:

2. $45 \div 9 =$ _____

3. $3 \times$ _____ $= 6$

4.

What will be the 25th figure in the pattern above?

5. $25 =$ _____ $\times$ _____

TUESDAY   Number Sense

1. How much money is four dollars, six quarters and two dimes?

2. How much money is eleven dollars, five nickels and six pennies?

3. Draw a combination of bills and coins for: $23.67

4. Draw a combination of bills and coins for: $89.11

5. If Spencer spent half of his $44.00 savings, how much did he spend?

WEDNESDAY — Geometry and Spatial Sense

1. How many right angles does a rectangle have?

2. How many faces does a rectangular prism have?

3. Are these shapes congruent or similar?

4. What shape is the face of a cylinder?

5. Draw a 90 degree angle.

THURSDAY — Measurement

1. What is the perimeter of the rectangle?

7 \
    13

2. What time is it?

3. 24.5 m = _____ cm

4. Draw a square with an area of 25 cm^2.

5. How many years in 4 decades?

Mary has 10 marbles in her pocket: 3 black, 2 red and 5 yellow.

1. What is the probability of picking a yellow marble?

2. What is the probability of picking a black marble?

3. What is the probability of choosing a red marble?

4. What is the probability of picking a marble?

5. What is the probability of choosing a blue marble?

BRAIN STRETCH

One orange costs $0.60, or you can buy a dozen for $5.99.

1. Which is the better price? Explain.

2. If you bought one dozen oranges and used a five dollar bill and a toonie to pay for your purchase, what would your change be?

MONDAY   Patterning and Algebra

1. Create a pattern:

   starts at 3, multiply by 2

2. 50 = _____ x 5

3. Extend the pattern.

   5, 6, 7, 5, 6, 7,_____, _____, _____

4. Are the sums equal?

   7 + 6   5 + 8

5. Create a shrinking pattern.

TUESDAY   Number Sense

1. What is the place value of 7
   in the number 8763?

2. What number comes after 2341?

3. What number comes before 908?

4. Madelyn is the eighth in line. How
   many people are in front of her?

5. Multiply:

   68
   x 7

WEDNESDAY — Geometry and Spatial Sense

1. What is a straight angle?

2. How many lines of symmetry?

H

3. How many faces does a pyramid have?

4. Draw a pair of intersecting lines.

5. Can you make a cube with this net?

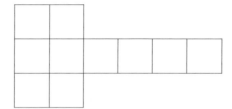

THURSDAY — Measurement

1. 400 dm = _____ km

2. What measuring tool would you use to measure the height of your room?

3. A cup of tea is about:
   - A. 1 L
   - B. 1ml
   - C. 250ml

4. 72 hours = _____days

5. How long might it take you to write your name?

Data Management

Here are the results of a Favourite Breakfast Food Survey.
Complete the chart and answer the questions about the results.

Favourite Breakfast Foods	Tally	Number
Cereal		17
Eggs		15
Pancakes		25
Grilled Cheese Sandwich		12

1. What was the most popular breakfast food? _____

2. What was the least popular breakfast food? _____

3. How many people liked either cereal or pancakes? _____

4. How many people liked pancakes more than eggs? _____

BRAIN STRETCH

Steve has a mass of 41 kg. Nick's mass is 44.5 kg and Tyler's mass is 38.2 kg.

1. How much heavier is Nick than Steve?

2. How much lighter is Tyler than Steve?

3. What is their total mass?

 Chalkboard Publishing © 2010

1. 45 x _____ = 90

2. 60 – 5 = _____

3.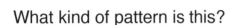

   What kind of pattern is this?

4. What will be the 20th figure?

5. Extend the pattern:

   1 1 2 2 3 3 4 4 1 1 2 2 3 _____, _____, _____, _____

TUESDAY Number Sense

1. Write a fraction equivalent to ½.

2. Write a fraction equivalent to ¾.

3. Which fraction is larger?

   ¼   ⅔

4. Which fraction is larger?

   ¾   ½

5. Which fraction is larger?

   ⅗   ½

WEDNESDAY — Geometry and Spatial Sense

1. A trapezoid has one pair of parallel sides.

   true   or   false

2. Draw a straight angle.

3. How many right angles does a trapezoid have?

   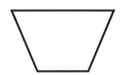

4. Are these shapes congruent or similar?

5. How many faces does a cylinder have?

THURSDAY — Measurement

1. Draw a rectangle that has a side 2 cm long.

2. Sunita can walk to school in 18 minutes. If she leaves at 8:10 am, when will she arrive at school?

3. Find the area of a square that has 3 cm sides.

4. 20 dm = ___ cm

5. How many years in a century? In a millennium?

Use the calendar to answer the questions.

June

Sunday	Monday	Tuesday	Wednesday	Thursday	Friday	Saturday
				1	2	3
4	5	6	7	8	9	10
11	12	13	14	15	16	17
18	19	20	21	22	23	24
25	26	27	28	29	30	

1. How many Saturdays are in the month of June? _____

2. What day of the week is June 9th? _____

3. Name the date that is 2 weeks after June 13th. _____

4. What is the date on the third Sunday in June? _____

5. What day of the week does the month end on? _____

BRAIN STRETCH

The Lemington Lions basketball team had a great season. They scored 26 points during their first game, 22 points during their second game, 33 points during their third game, 40 points during their fourth game and 34 points during their last game.

1. How many points did they score all season?

2. What is the range of their points scored?

MONDAY — Patterning and Algebra

1. 25 x ___ = 225

2. Extend the pattern:

   2, 5, 8, 11, ____ , ____ , ____

3. Are the sums equal?

   11 + 14   19 + 8

4. Find the first four numbers of the pattern:

   start at 7, add 4

5. Extend the pattern:

   4, 7, 10, 13, ____ , ____ , ____

TUESDAY — Number Sense

1. Name 3 composite numbers.

2. Name 3 prime numbers.

3. Multiply: 512
                 x 3

4. Multiply: 19 x 8

5. Megan has two sets of hockey cards. Each set has nine cards. How many cards does she have altogether?

WEDNESDAY — Geometry and Spatial Sense

1. Draw an obtuse angle.

2. Draw an acute angle.

3. How many lines of symmetry does the following letter have?

   ## U

4.

   measure of angle _____

   type of triangle _____

5. Reflect this shape

THURSDAY — Measurement

1. 42 days = _____ weeks

2. Michael went to do his homework at 8:15 pm and finished at 9:18 pm. How long did it take?

3. Compare the following using: >, < or =

   18 cm ☐ 180 mm

4. How long does it take to tie your shoes:

   1 hour   1 min   1 day

5. Find the perimeter of this shape.

   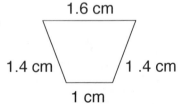

Ben went fishing. Look at the chart to see the number of fish Ben caught over a Monday to a Friday.

Day of the Week	Monday	Tuesday	Wednesday	Thursday	Friday
The Number of Fish Caught	5	10	15	20	25

1. On what day did Ben catch the most number of fish? _____

2. On what day did Ben catch the least number of fish? _____

3. How many fish did Ben catch on Tuesday and Thursday? _____

4. What is the difference between the most number of fish Ben caught and the least number of fish? _____

BRAIN STRETCH

Can you name three 3D shapes that can roll? Draw them.

MONDAY — Patterning and Algebra

1. Create a shrinking pattern.

2. Make the statement true:

$$30 + \underline{\quad} = \underline{\quad} + 10$$

3. Are these differences equal?

    30 - 18   20 - 6

4. Extend the pattern:

    144, 134, 124, 114, _____, _____, _____

5. What is the pattern rule for question number 4?

TUESDAY — Number Sense

1. Divide:

    $42 \div 7 =$

2. Divide:

    $62.9 \div 10 =$

3. Multiply:

    $5 \times 9 =$

4. Divide:

    $6\overline{)420}$

5. A basket holds eight pears. How many pears do six baskets hold?

WEDNESDAY — Geometry and Spatial Sense

1. How many faces does a cube have?

2. Are these shapes similar or congruent?

3. How many inside obtuse angles does a pentagon have?

4. What 3D figure can be made with these pieces?

5. Draw a pair of congruent shapes.

THURSDAY — Measurement

1. What measuring tool would you use to find the mass of your body?

2. Find the area of this rectangle:

   6 cm

   3.6 cm

3. A bucket can hold about:

   A. 6 L of water

   B. 6 drops of water

   C. 6 ml of water

4. Katelyn took 3 hours and 20 minutes to finish reading a book. How many minutes was that in total?

5. 2 years = _____ weeks

Here are the results of a Favourite Sport Survey.
Complete the chart and answer the questions about the results.

Snack	Tally	Number
Basketball		18
Baseball		9
Football		7
Soccer		15

1. What was the most popular sport? _____

2. What was the least popular sport? _____

3. How many people liked either soccer or football?_____

BRAIN STRETCH

Madelyn practices the violin for 6.5 hours a week.

1. How many hours a month does she practice?

2. How many hours a year does she practice?

1. Fill in the missing number:

   6,9,12,_____,18,21,24

2. 60 = 25 +_____

3. Complete the pattern:

   1,2,4,8,16, _____,

4. Find the first five numbers of the pattern rule:

   start at 17, add 3

5. Write a number sentence that equals 81 ÷ 9 = 9.

1. Multiply:

   23.6 X100

2. Divide:

   56 ÷ 8 =

3. Divide:

   261.8 ÷ 100 =

4. A horse has four legs. How many legs do eleven horses have?

5. Subtract: $94.87
             - $32.38

1. Classify the angle.

2. Classify the angle.

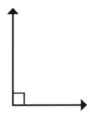

3. Draw the reflection:

4. Draw a net for a pyramid.

5. How many edges does a rectangular prism have?

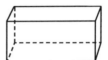

1. Find the area of this shape.

4

7

2. A bicycle is likely:

A. 20 kg

B. 20 g

C. 20 mg

3. 60 years = _____ decades

4. Liam walked 1.4 km to the beach. How many metres did he walk?

5. Draw a line that is 45mm long.

FRIDAY — Data Management

Mr. Wellars computer class recorded their typing speeds. Here is a list of how many words per minute they can type:

12	16	11	12	14	20
11	19	17	11	17	12
16	18	12	19	19	12

1. What is the range of the data?

2. What is the mode?

3. What is the fastest speed in the class?

4. What is the slowest speed in the class?

5. How many students are in the class?

BRAIN STRETCH

Jennifer, Laura and Karen were buying supplies for a school project. They bought a sheet of Bristol board for $1.75, a glue stick for $2.29, markers for $3.55 and construction paper for $1.99.

1. How much did they spend altogether?

2. How much will each girl have to spend?

1. Are the quotients equal?

   $36 \div 4$ $28 \div 7$

2. Complete the pattern:

   225, 230, 235, 240,_____, _____, _____

3. Complete the pattern:

   109, 105, 101, 97, _____, _____, _____

4. $80 - 61 =$

5. What is the pattern rule?

   1000, 900, 800, 700, 600

TUESDAY   Number Sense

1. A bicycle has two wheels. How many wheels do six bikes have?

2. Madelyn is 9th in line to get a drink after lunch. How many people are in front of her?

2. Multiply:

   $4 \times 6 =$

3. Gum is sold in packs of eight pieces. How many pieces are in three packs?

4. Name three odd numbers between 50 and 100.

WEDNESDAY · Geometry and Spatial Sense

1. Name two quadrilaterals.

2. How many faces does a rectangular prism have?

3. Classify the angle.

4. How many acute angles does a trapezoid have?

5.

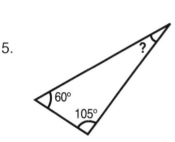

measure of angle _____

type of triangle_____

THURSDAY · Measurement

1. Complete the following using >,< or =:

   56 km ☐ 5000 m

2. Find the perimeter of the octagon.

   12

3. John delivers newspapers after dinner for 3.5 hours. If he starts at 6:15 pm, when will he finish?

4. What time is it?

5. 80 minutes =_____hours_____minutes

FRIDAY — Data Management

Lawrence surveyed kids in his grade to find out what type of books they liked to read. Here are his results:

Fiction 25   **Non-Fiction 12**   **Science-Fiction 5**   **Comic Books 22**   **Joke Books 26**

Create a tally chart to show the results.

Fiction	Non-fiction	Science Fiction	Comic Books	Joke Books

1. What is the most popular type of book?   2. What is the least popular type of book?

3. What are the top three types of books?   4. How many people did Lawrence ask?

5. What is the range of the results?

BRAIN STRETCH

Two times a number, plus 15, is 31. What is the number?

MONDAY — Patterning and Algebra

1. What is the missing number?

   _____ $\div$ 21 = 7

2. 180 - _____ = 65

3. What will be the 16th shape in the pattern?

5. What kind of pattern is this?

   1223334444455555

4. Complete the pattern:

   3, 10, 17, 24, ___, ___, ___

TUESDAY — Number Sense

1. Divide:

   $36 \div 9 =$

2. Round to the nearest thousand

   8725

3. Write the equivalent fraction.

   $\dfrac{3}{10} = \dfrac{}{20}$

4. Choose >, <, or =.

   2122 ☐ 1221

5. Madelyn bought 5 bunches of tulips. Each bunch cost $3.20. She paid with $20.00. What was Madelyn's change?

WEDNESDAY — Geometry and Spatial Sense

1. Can a cube be made from this net?

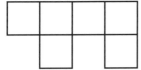

2. Draw a pair of similar shapes.

3. How many triangular faces does a triangular prism have?

4. How many lines of symmetry?

P

5. Which pair of lines are not parallel?

A. 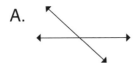   B.   C.

THURSDAY — Measurement

1. 220 mm = _____ cm

2. What measuring device would you use to find a cup of milk?

3. Megan's dog weighs 10 kg. How many grams is that?

4. 2100 ml = _____ L

5. Compare the following using: >, < or =

120 mins [   ] 1.5 hours

Chalkboard Publishing © 2010

Stacy worked at The Olde Ice Cream Shoppe in the summer. She kept a tally of all the sales she made and created this pictograph from her results.

Chocolate	🍦🍦🍦🍦🍦🍦🍦
Vanilla	🍦🍦🍦🍦🍦
Butterscotch	🍦🍦🍦🍦🍦🍦
Strawberry	🍦🍦🍦
Maple Walnut	🍦
Bubble Gum	🍦🍦🍦🍦🍦

 = 2 ice cream cones

1. What was the most popular flavour?

2. What was the least popular flavour?

3. Which two flavours sold the same amount?

4. How many purchases were made on the day the survey was taken?

5. What is the range of the data?

BRAIN STRETCH

How much money in total?

3 twenty-dollar bills, 1 toonie, 3 loonies, 3 quarters, 4 dimes, and 9 nickels

MONDAY Patterning and Algebra

1. Write the following in another way.

   8 + 8 + 8 + 8

2. Are the products equal?

   7 x 6   4 x 12

3. _____ x 6 = 54

4. What is the pattern rule?

   900, 850, 800, 750

5. What kind of pattern is this?

   105, 115 ,125, 135, 145, 155,

TUESDAY Number Sense

1. Divide:

   $4\overline{)29}$

2. Divide:

   $3\overline{)35}$

3. How many quarters are in 10 loonies?

4. Subtract:

   145 − 91 =

5. Shelly had 27 shells. She put the shells in groups of 5. How many equal groups of shells were there?

WEDNESDAY Geometry and Spatial Sense

1. Name a polygon with 5 vertices.

2. Which shape has 2 pairs of parallel lines and all sides are equal?

3. Draw a 180 degree angle.

4. Can a cone be made from this net?

5. This is a _____.

How many?   faces _____   edges_____   vertices _____

THURSDAY Measurement

1. Would you measure the contents of a bathtub using ml or L?

2. What temperature is it most likely to be if you are outside playing in the snow?

   A.   − 5°C

   B.   30°C

   C.   4°C

3. Spencer brushes his teeth twice a day for 2 minutes. How many minutes does he brush his teeth each week?

4. How many decades in 5 centuries?

5. It is 4:00 pm. What time will it be in 1 hour and 14 minutes?

Emily's mom brought a bag of fruit to the family picnic. She put: 3 bananas, 5 apples, 10 apricots and 2 mangoes in the bag. What is the probability of randomly selecting the following?

What is the probability as a fraction of:

1. a banana

2. a mango

3. an apple

4. an apricot or mango

5. a chocolate bar

BRAIN STRETCH

How many of these nets can be made into cubes?

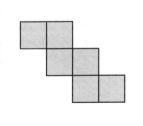 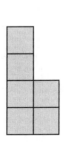 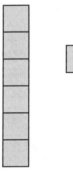 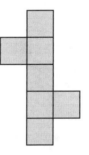 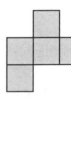 

 Week 22

MONDAY — Patterning and Algebra

1. $45 \div 5 = 21-$ _____

2. Extend the pattern:

   55, 61, 67, 73, ___, ___, ___

3. Quadrilaterals are shapes with 4 sides.

   true   or   false

4. Find the first four numbers of the pattern rule:

   start at 202, add 6

5. What will be the 12th number in this pattern?

   4, 8, 12, 16, 20, 24

TUESDAY — Number Sense

1. Divide:

   $106 \div 2 =$

2. What is 124 divided by 4?

3. There were 30 children. 19 of the children were boys. What fraction of the children were girls?

4. What is 81 divided by 3?

5. Vicki wants to share her twenty jelly beans equally with her best friend. How many can each have?

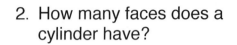

1. Is a 120 degree angle obtuse, acute or right?

2. How many faces does a cylinder have?

3. Draw a reflection of this shape:

4. How many triangular faces does a cube have?

5. Classify the following pair of lines.

    A. intersecting   B. parallel C. perpendicular

THURSDAY Measurement

1. 3305 m = _____ km

2. Draw a rectangle with an area of 18 units².

3. 6 months = _____ weeks

4. The best estimate of the weight of Ben's frog is:

    A. 50 g

    B. 5 g

    C. 1500 g

5. Spencer can hop on one foot for 200 seconds. How many minutes is that?

Here is a pie graph to show students' favourite seasons.

Students' Favourite Season

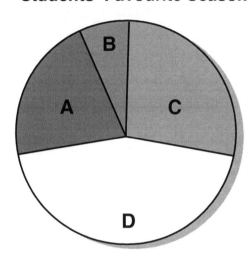

A - Spring
B - Fall
C - Winter
D - Summer

1. Which season was picked the most?

2. Which season was picked the least?

3. Which two seasons were picked about the same?

BRAIN STRETCH

A dozen eggs cost $1.20. How much do 10 dozen eggs cost?

1. Extend the pattern:

   100, 99, 97, 94, _____, _____, _____

2. What is the next shape in the pattern?

   ⟹ ⟸ ⟹ ⟸

3. Are the products equal?

   3 x 8   4 x 6

4. What is the pattern rule?

   99, 93, 87, 81, 75

5. 55 x _____ = 110

TUESDAY — Number Sense

1. Susan bought three skirts for $9.99 each. How much was the total bill?

2. Add:

   2778
   +4619

3. Subtract:

   81 730
   - 14 458

4. Chris gave 630 hockey cards to his friends. He gave the same number of cards to each of his 10 friends. How many cards did each friend get?

5. What is the difference between 8.4 and 2.6?

WEDNESDAY — Geometry and Spatial Sense

1. Show a line of symmetry on this shape:

2. Can a cube be made from this net?

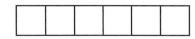

3. Are these shapes similar or congruent?

4. How many right angles can a triangle have?

5. Look at the shapes. Choose flip, slide, or turn.

THURSDAY — Measurement

1. Kristine's family is going on a vacation for 3 weeks. How many days is that?

2. Find the perimeter of a pentagon whose sides are 2.2 m long.

3. What measuring tool would you use to find the width of a car?

4.

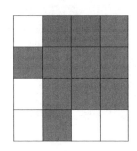

Perimeter = _____ units

Area = _____ square units

5. 30 km = _____ m

Mrs. Carter's class regularly exercises to stay fit. They counted how many push ups they can do in a minute. Here are their results:

11	5	16	28	20	11
8	8	15	11	9	15
20	24	21	11	19	12

1. What is the range of the data collected?

2. What is the mode?

3. How many students were surveyed?

4. What is the greatest number of push ups done in one minute?

5. What is the least number of push ups done in a minute?

BRAIN STRETCH

Michael bought 4 boxes of holiday cards. Each box has 18 cards.

1. How many cards were there altogether?

2. If he paid $24.96, how much was each box?

Patterning and Algebra

1. Are these differences equal?

    65 - 12   77 - 21

2. Make this statement true:

    33 + ___ = 50 - 10

3. Complete the pattern:

    5, 13, 21, 29, ____, ____, ____

4. What is the missing number?

    [ ] x 100 = 900

5. Show the first five numbers of this pattern:

    start at 60, subtract 7

TUESDAY Number Sense

1. Arrange these digits to make the smallest possible number.

    6 9 2 4

2. Write $43.89 in words.

3. What is the following decimal as a fraction?

    0.67

4. What is the sum of 4.2 and 5.9?

5. Sam has eighty-one trading cards. He has twelve more than John. How many cards does John have?

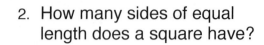

WEDNESDAY — Geometry and Spatial Sense

1. Draw two lines of symmetry on this shape.

2. How many sides of equal length does a square have?

3. Draw a set of parallel lines.

4. How many vertices does this shape have?

5. A soda can is what kind of 3D figure?

    A. cube

    B. cone

    C. cylinder

THURSDAY — Measurement

1. 2 years = _____ days

2. Find the area of this shape:

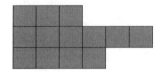

3. Draw a worm that is 25 mm long.

4. Compare the following using: >,< or =

    11 minutes ☐ 200 seconds

5. Draw a square with a perimeter of 24 units.

Data Management

Students at Orchard Park Public School can choose any event to participate in during the school's annual Spring Field Day. Here are some of the choices:

Standing Long Jump	45 students
High Jump	14 students
Shot Put	31 students
Running Long Jump	38 students
100 m Sprint	60 students
1500 m Run	38 students

1. How many students were surveyed?

2. What was the most popular event?

3. What was the least popular event?

4. How many students chose either shot put or 100 m sprint?

5. What is the range of the data?

BRAIN STRETCH

Sophie's favourite CD has 10 songs on it. Its total length is 39 minutes. On average, how many minutes was each song?

MONDAY — Patterning and Algebra

1. What is the pattern rule?

   60, 75, 90, 105, 120

2. Are the products equal?

   6 x 8   11 x 4

3. In which number sentence does a 10 make the equation true?

   A.   8 x ___ = 48

   B.   18 ÷ ___ = 2

   C.   190 + ___ = 200

4. Create a growing pattern.

5. 100 = 25 + 20 + ____

TUESDAY — Number Sense

1. Divide:

   64 ÷ 8 =

2. Draw the least amount of coins needed to represent $7.71.

3. What is $\frac{89}{100}$ as a decimal?

4. Multiply:

   3.9 x 100 =

5. Mary got $12.88 back in change when she bought her markers. She paid with a twenty dollar bill. How much did the markers cost?

1. How many edges does a cylinder have?

2. Draw an acute angle.

3. Draw a straight angle.

4. Draw two lines of symmetry on this shape:

5. How are a circle and triangle different?

1. Compare the following using: >,< or =

2 kg ☐ 1098 g

2. Lisa carries a lot of books home from school. Her backpack most likely weighs about:

    A. 10 kg

    B. 10 g

    C. 0 mg

3. Robin's birthday is six days after Valentine's Day. When is her birthday?

4. 6 dm =_____mm

5. 2680 g = _____ kg

Here is a bar graph of how much money the grade 4 class fund raised for a class trip.

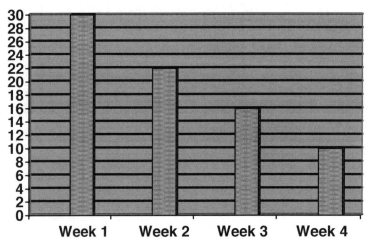

Trip Fund Raising

1. What week was the most money raised? _____

2. How much money was raised in week 2? _____

3. How much money was raised altogether? _____

4. How much more money was raised in week 3 than in week 4? _____

BRAIN STRETCH

Farmer Felipe has a large corn field measuring 30 m by 24 m. He wants to build a fence around it.

1. How many square metres of corn can he plant?

2. What is the perimeter of the field?

3. If fencing costs $8 per metre, how much will his fence cost altogether?

MONDAY Patterning and Algebra

1. What are the first six numbers of this pattern?

   start at 220, subtract 11

2. Fill in the blank to make the statement true.

   $5 \times 12 = 70 - \_\_\_$

3. In which number sentence does a 7 make the equation true?

   A. $3 \times \_\_\_ = 18$

   B. $28 \div \_\_\_ = 4$

   C. $24 + \_\_\_ = 35$

4. What is the pattern rule?

   2, 13, 24, 35, 46, 57

5. Complete the pattern:

   89, 88, 87, 89, ____, ____, ____

TUESDAY Number Sense

1. Order these numbers from greatest to least:

   8901, 9801, 8109

   _____ , _____ , _____

2. Choose >, <, or =.

   98 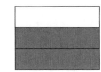 89

3. Round 2398 to the nearest thousand.

4. What fraction is shaded?

5. Add:

   $\dfrac{4}{8} + \dfrac{3}{8} =$

WEDNESDAY — Geometry and Spatial Sense

1. How many lines of symmetry?

   E

2. Is a 91 degree angle obtuse, acute or right?

3. How many faces does a sphere have?

4. Draw a line of symmetry on this shape:

5. Draw the faces of a triangular based pyramid.

THURSDAY — Measurement

1. What is the area of a square that has 7 dm sides?

2. Beatrice bathes four times a week. How many times does she bathe in a year?

3. 360 seconds = _____ minutes

4. What measuring tool would you use to find the depth of a kitchen sink?

5. Draw a square with a perimeter of 20 units.

Henry surveyed his neighbours to see which newspapers they subscribed to. Here are his results:

Telegram	18
Mirror	25
Daily	9
None	4

1. Create a pictograph below using the data in the space above. What symbol will you use to represent a paper? What key will you use?

BRAIN STRETCH

Divide a number by 8, subtract 2, and multiply it by 4. The result is 32. What is the number?

MONDAY — Patterning and Algebra

1. Fill in the blank to make the statement true.

   $2 \times 12 =$ _____ $\times 6$

2. Complete the pattern:

   67, 65 , 63, 61, _____, _____, _____

3. Are the quotients equal?

   $54 \div 9$ $49 \div 7$

4. What will be the 8th number in this pattern?

   7, 14, 21, 28, 35

5. In which number sentence does a 9 make the equation true?

   A. $3 \times$ ____ $= 24$

   B. $18 \div$ ____ $= 2$

   C. $24 +$ ____ $= 30$

TUESDAY — Number Sense

1. Subtract;

   $\dfrac{4}{6} - \dfrac{1}{6} =$

2. Add:

   $398 + 125 =$

3. Write 2745 in words.

4. How many legs do six spiders have?

5. Divide:

   $4\overline{)28}$

Chalkboard Publishing © 2010

WEDNESDAY · Geometry and Spatial Sense

1. How many right angles does a square have?

2. Reflect this shape.

3. How many vertices does a sphere have?

4. How many angles are in a circle?

5. Draw an obtuse angle.

THURSDAY · Measurement

1. What would be the best estimate of how many litres of water could fit in a pitcher?

   A. 1 L

   B. 10 L

   C. 100 L

2. Which has a greater mass: a horse or a dog?

3. What is the perimeter of a hexagon with 5 m sides?

4. 410 cm =_____ dm

5. How many days in 6 years?

Mr. Lewis posted the class marks for the science test. Here are the results:

88	90	76	30	90
100	100	88	90	20
80	72	100	88	100
48	100	86	76	98

1. What is the highest mark?

2. What is the lowest mark?

3. What is the mode?

4. How many students wrote the test?

5. What is the range of the marks?

BRAIN STRETCH

Gerry ordered eight pizzas for his birthday party. Each pizza had ten slices.

1. How many pieces of pizza will there be altogether?

2. If there are 18 people at his party, how many pieces of pizza can each person have?

1. Create a repeating pattern.

2. Show the first 3 numbers in this pattern:

   start at 200, subtract 15

3. In which number sentence does 4 make the equation true?

   A. $20 \times \underline{\quad} = 80$

   B. $18 \div \underline{\quad} = 2$

   C. $24 + \underline{\quad} = 30$

4. What will be 14th number in this pattern?

   54, 51, 48, 45, 42, 39

5. Fill in the blank to make the equation true.

   $40 - 9 = \underline{\qquad} + 23$

TUESDAY — Number Sense

1. Divide:

   $128.9 \div 1000 =$

2. Write six tenths in numbers.

3. Order these numbers from least to greatest.

   345, 545, 435, 334

   _____, _____, _____, _____

4. 8 novels cost $40. How much does each novel cost?

5. Name the factors for 72.

1. A scalene triangle has 3 equal sides.

    true   or   false

2. Draw a figure that does not have a line of symmetry.

3. Draw a trapezoid.

4. Are these shapes similar?

5. Draw a pair of congruent shapes.

THURSDAY — Measurement

1. Find the perimeter of a hexagon with 4 mm equal sides.

2. 96 hours = _____ days

3. Draw a rectangle with an area of 72 square units.

4. Draw a letter N that is 3 cm tall.

5. Connor has a 30 minute piano lesson each week. How many hours of lessons does he have in six months?

FRIDAY — Data Management

Here is a graph of the most popular types of movies this holiday:

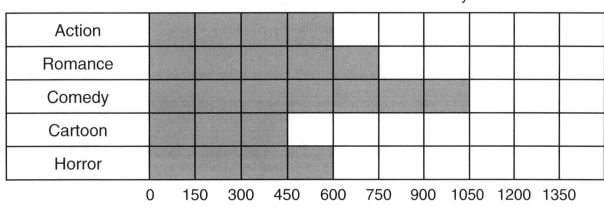

Movies Seen This Holiday

Type of Movie (vertical axis)

Type of Movie	0	150	300	450	600	750	900	1050	1200	1350
Action	▓	▓	▓	▓						
Romance	▓	▓	▓	▓	▓					
Comedy	▓	▓	▓	▓	▓	▓	▓			
Cartoon	▓	▓	▓							
Horror	▓	▓	▓	▓						

Number of People

1. What is the range of the data?

2. What is the most popular type of movie?

3. What is the median?

4. What two types of movies did people like the same?

5. How many people went to the theatre altogether during the holiday?

BRAIN STRETCH

David and Demetra went for a 4 km bike ride. How many metres did they ride?

MONDAY — Patterning and Algebra

1. What is the missing number?

   _____ ÷ 12 = 144

2. Are the products equal?

   3 x 9   4 x 8

3.   66 - _____ = 30 + 20

4. What is the pattern rule?

   32, 36, 40, 44, 48, 52

5. John runs 3 km every day. How many km will John run in 45 days?

TUESDAY — Number Sense

1. Name the factors for 48.

2. Divide:

   $6\overline{)38}$

3. Write 67 523 in expanded form.

4. What number is 10 less than 48 352?

5. Multiply:

   99 x 100 =

1. Can you make a pyramid with this net?

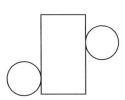

2. An equilateral triangle has 3 equal sides.

   true   or   false

3. How many edges does a rectangular prism have?

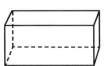

4. Draw the reflection of this shape:

5. What does congruent mean?

THURSDAY | Measurement

1. 11 cm =_____m

2. Compare the following using: >, < or =

   3.5 [ ] 5.3

3. Rob plays video games for 20 minutes each day. How long does he play in four weeks?

4. What is the area of this rectangle?

   8m
   1m [   ]

5. What measuring device would you use to find the length of a movie?

Answer the probability questions using the information on the two spinners.

Spinner 1

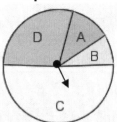

Spinner 2

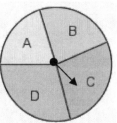

If someone has only 1 spin:

1. Is spinning C more probable on Spinner 1 or Spinner 2? _____

2. On which spinner is A more probable? _____

3. On which sipnner is B less probable? _____

4. Which letter is equally likely on both Spinner 1 and Spinner 2? _____

BRAIN STRETCH

Elizabeth's hockey team won 12 games and lost 3 games this season.

1. How many games did they play altogether?

2. If Elizabeth's mother had to pay $3 to watch her games, how much did she have to pay for the season?

3. Her hockey team will play three times as many games next year. How many games will that be?

Math — Show What You Know!

_____

☐ I read the question and I know what I need to find.

☐ I drew a picture or a diagram to help solve the question.

☐ I showed all the steps in solving the question.

☐ I used math language to explain my thinking.

Student Tracking Sheet

Student	Week 1	Week 2	Week 3	Week 4	Week 5	Week 6	Week 7	Week 8	Week 9	Week 10	Week 11	Week 12	Week 13	Week 14	Week 15

Student Tracking Sheet

Student	Week 16	Week 17	Week 18	Week 19	Week 20	Week 21	Week 22	Week 23	Week 24	Week 25	Week 26	Week 27	Week 28	Week 29	Week 30

You Are Incredible!

Keep Up the Good Work!

_____

Week 1

Mon. **1.** 14 **2.** 7 **3.** 15, 18, 21 **4.** shrinking **5.** 10, 15, 20, 25

Tues. **1.** 6 **2.** 7911, 7918, 8791 **3.** 6531 **4.** 460, 280 **5.** 40

Wed. **1.** trapezoid or quadrilateral **2.** 3 or more **3.** cone **4.** turn
5. a 4 sided figure/polygon

Thurs. **1.** 10mm **2.** 4:45 **3.** cm **4.** kilometre **5.** 18, 16

Fri. **1.** 42 **2.** hockey **3.** 2 **4.** soccer, basketball, hockey **5.** ans. will vary

Brain Stretch 52, 104, 260

Week 2

Mon. **1.** 15 **2.** 25, 30, 35 **3.** 11, 22, 11 **4.** 40, 30, 20 **5.** 9

Tues. **1.** 12 205 **2.** 4723, 7324, 7432 **3.** 6541 **4.** 1300, 2700 **5.** $28.55

Wed. **1.** pentagon **2.** 2 sets/pairs **3.** cylinder **4.** flip **5.** ans. vary

Thurs. **1.** 1000 **2.** a **3.** a **4.** 3:35 **5.** m

Fri. **1.** reggae ||||| ||||| || classical || dance ||||| ||||| ||||| ||||| ||||| ||| rock ||||| ||||| ||||| ||
country ||||| ||||| |||| **2.** dance music **3.** classical **4.** 14 **5.** 73

Brain Stretch 634

Week 3

Mon. **1.** 25 **2.** growing **3.** 22 **4.** 8 **5.** 55, 66, 77

Tues. **1.** 8983 **2.** 5 **3.** 650 **4.** 816 **5.** $26.28

Wed. **1.** right triangle **2.** 2 sets/pairs **3.** square prism or cube **4.** a **5.** 8, 8

Thurs. **1.** 100 **2.** 20 **3.** m or km **4.** 1:40 **5.** 10

Fri. **1.** title- answers will vary X-axis number of votes **2.** blue **3.** green, yellow **4.** 36

Brain Stretch ans. will vary

Week 4

Mon. **1.** growing **2.** growing and repeating **3.** decreasing **4.** 35 **5.** 11

Tues. **1.** 2543, 2745, 3245 **2.** 5028, 5208, 8052 **3.** 3353 **4.** 180 **5.** 8621

Wed. **1.** parallelogram or quadrilateral **2.** octagon **3.** 0 **4.** 3 **5.** square based pyramid

Thurs. **1.** 60 **2.** 7 **3.** mm or cm **4.** 8:30 **5.** 16 cm, 12 cm^2

Fri. **1.** Oak Town **2.** Birch Town **3.** 80 **4.** 60 mm **5.** Birch, Pine, Maple,
Spruce, Elm, Oak

Brain Stretch 14 m = 1400 cm

Week 5

Mon. **1.** 2, 4, 6, 8 **2.** 1, 6, 11, 16 **3.** 20, 18, 16, 14 **4.** 3, 6, 9, 12 **5.** 8

Tues. **1.** 9419 **2.** 5280 **3.** 5821 **4.** = **5.** 7431

Wed. **1.** octagon **2.** pentagon **3.** 1 **4.** 5 **5.** cube or square prism

Thurs. **1.** 30 **2.** 48 m **3.** 9 **4.** 9:55 **5.** 50 m

Fri. cars – 11 trucks – 4 vans – 13 buses – 22 **1.** 50 **2.** buses **3.** 2 **4.** 18 **5.** 18

Brain Stretch 3 computers

Week 6

Mon. **1.** 30, 35, 40, 45 **2.** 100, 80, 60, 40 **3.** 9 **4.** start at 8, multiply by 2 **5.** 7

Tues. **1.** (200+ 100=300) ans. may vary **2.** 900 +500=1400 (ans. may vary) **3.** 479 **4.** 787 **5.** 381

Wed. **1.** a point where 2 sides of a figure meet **2.** 4 **3.** **4.** sphere **5.**

Thurs. **1.** ans. will vary −3 by 8, 4 by 6 **2.** 28 units^2 **3.** ans. will vary −
1 by 5, 2 by 4, 3 by 3 **4.** 8:55 **5.** 240 mins

Fri. **1.** cats **2.** 75 **3.** 14 **4.** 25

Brain Stretch 3.5h per week 15 hrs. per month (based on a 30 day month)

Week 7

Mon. **1.** 7 **2.** 1 **3.** 30 **4.** 68 **5.** ans. will vary

Tues. **1.** 80 **2.** 200 **3.** 114 **4.** 121 **5.** $54.00

Wed. **1.** **2.** 5 sides **3.** no **4.** all except triangle **5.** circle

Thurs. **1.** ans. will vary **2.** 8000 **3.** mL **4.** 3:20 p.m. **5.** 32 cm 63 cm^2

Fri. **1.** not likely **2.** very likely **3.** 3/10 **4.** 7/10 **5.** ans. may vary

Brain Stretch **1.** 50 m **2.** 150 m2 **3.** $500.00

Week 8

Mon. **1.** 4, 16, 64, 256 **2.** 2, 6, 18, 54 **3.** 49 **4.** start at 99, subtract 11 **5.** ans. will vary

Tues. **1.** 6039 **2.** 1004 **3.** 40 **4.** 19 **5.** 62

Wed. **1.** false **2.** c **3.** 8 **4.** ans. will vary **5.** first pair

Thurs. **1.** 20 cm **2.** 200 000 **3.** c **4.** 300 **5.** 12 units^2

Fri. **1.** 1/6 **2.** 1/6 **3.** 3/6 or ½. 3/6 or ½ **4.** impossible **5.** each tally mark =1, = 34

Brain Stretch **1.** .5 L **2.** 13 jars **3.** $26.00

Week 9

Mon. **1.** 81, 243, 729 **2.** 20, 24, 28 **3.** 24, 30, 36 **4.** 12, 10, 8 **5.** ans. will vary

Tues. **1.** 1, 2, 3, 4, 6, 8, 12, 24 **2.** 2 R3 **3.** 40 000+5000+20+3 **4.** 2512 **5.** 2465

Wed. **1.** b **2.** isosceles **3.** a **4.** acute **5.** cylinder, cone, sphere

Thurs. **1.** 2 **2.** 16000 **3.** 17.2 cm **4.** 30 **5.** 9 m^2

Fri. **1.** circle or pie **2.** amusement park **3.** swim, shop and eat, go to amusement park
4. sleeping **5.** ans. will vary (Disney World, Florida)

Brain Stretch oldest – Vivienne youngest - Tina

Week 10

Mon. **1.** yes **2.** 9 **3.** ans. will vary **4.** a **5.** 90

Tues. **1.** 9 **2.** 9832 **3.** 1048 **4.** 7450, 4879, 4531, 1751 **5.** $47.81

Wed. **1.** 0 **2.** rectangle **3.** 5 **4.** rectangular prism **5.** isosceles, equilateral, scalene, right

Thurs. **1.** 11:10 **2.** 6000 m **3.** 40 units **4.** 1984 **5.** 3

Fri. **1.** all except Josie **2.** Martin, Darcy, Martha, Michael **3.** all **4.** 37

Brain Stretch all together = 5.3 km 0 .7 km farther

Week 11

Mon.   **1.** 34, 44, 54 **2.** 160, 320, 640 **3.** 72 **4.** 150 **5.** ans. will vary

Tues.   **1.** 7/8, 4/8, 2/8 **2.** 3/5, 2/5, 1/5 **3.** 2/7 **4.** 4 **5.** 3 loonies, 2 dimes

Wed.   **1.** 2 **2.** hexagon **3.** parallelogram, square, rectangle **4.** none **5.** 1

Thurs.   **1.** 11:45 **2.** 3:10 p.m. **3.** 26 cm **4.** 1944 **5.** a- 4 mins

Fri.   **1.** 170 **2.** snowboarding **3.** ice fishing **4.** 40 **5.** 90

Brain Stretch   each day = 71   in a week = 497

Week 12

Mon.   **1.** ans. will vary **2.** 4 **3.** 55, 66, 77 **4.** 12, 28 **5.** 3

Tues.   **1.** 1/5, 3/5, 4/5 **2.** .4 **3.** 57/100 **4.** 3/7 **5.** $67.80

Wed.   **1.** obtuse **2.** 0 **3.** 4 **4.** a closed figure with 3 or more sides **5.** AB - CD

Thurs.   **1.** 11:40 **2.** 5:00 p.m. **3.** 18, 16 **4.** 8hrs. 50mins. **5.** 600 000

Fri.   ans. will vary for all

Brain Stretch   **1.** 99 m   **2.** 432   **3.** 4:00 p.m.

Week 13

Mon.   **1.** ans. will vary **2.** 5, 5, 5, 5, 5 **3.** 15 **4.** yes **5.** 44

Tues.   **1.** $90.00 **2.** $480.00 **3.** ans. will vary **4.** $35.00 **5.** $18.56

Wed.   **1.** 0 **2.** parallel lines **3.** ans. will vary **4.** equilateral **5.** slide

Thurs.   **1.** 9 units^2 **2.** 12:45 **3.** check ans. with a ruler **4.** a **5.** 1000

Fri.   **1.** 3 **2.** 2 **3.** 5 **4.** 2

Brain Stretch   **1.** $130.00 **2.** between 11:00 a.m. and 11:30 a.m. **3.** $4.75

Week 14

Mon.   **1.** ans. will vary **2.** 5 **3.** 2 **4.** ice cream **5.** 5x5 or 1x25

Tues.   **1.** $5.70 **2.** $11.31 **3.** ans. will vary **4.** ans. will vary **5.** $22.00

Wed.   **1.** 4 **2.** 6 **3.** similar **4.** circle **5.** └─

Thurs.   **1.**   **2.** 6:05 **3.** 2450 **4.** 5 cm x 5 cm **5.** 40

Fri.   **1.** 5/10 or 1/2 **2.** 3/10 **3.** 2/10 or 1/5 **4.** 10/10 **5.** 0/10

Brain Stretch   **1.** It's better to buy a dozen because if you bought 12 individually it would cost $7.20. By buying the dozen at $5.99 you will save $1.21. **2.** $1.01

Week 15

Mon.   **1.** 3, 6, 12, 24 **2.** 10 **3.** 5, 6, 7 **4.** yes **5.** ans. will vary

Tues.   **1.** hundreds **2.** 2342 **3.** 907 **4.** 7 **5.** 476

Wed.   **1.** 180 degrees **2.** 2 **3.** ans. will vary depending on the base **4.** ans. will vary **5.** no

Thurs.   **1.** .04 **2.** metre stick **3.** c **4.** 3 **5.** 5-10 seconds (will vary)

Fri.   **1.** pancakes **2.** grilled cheese **3.** 42 **4.** 10

Brain Stretch   **1.** 3.5 kg **2.** 2.8 kg **3.** 123.7 kg

Week 16

Mon.   **1.** 2   **2.** 55   **3.** repeating   **4.** oval   **5.** 3, 44, 11, 22

Tues.   **1.** ans. will vary – 2/4, 3/6   **2.** ans. will vary – 6/8   **3.** 2/3   **4.** 3/4   **5.** 3/5

Wed.   **1.** true   **2.** ←⌣→   **3.** 0   **4.** congruent   **5.** 2

Thurs.   **1.** measure for accuracy   **2.** 8:28 a.m.   **3.** 9 cm^2 **4.** 200   **5.** century = 100 yrs, millennium = 1000 years

Fri.   **1.** 4   **2.** Friday   **3.** 27th **4.** 18th **5.** Friday

Brain Stretch   **1.** 155   **2.** 18

Week 17

Mon.   **1.** 9   **2.** 14, 17, 20   **3.** no   **4.** 7, 11, 15, 19   **5.** 16, 19, 22

Tues.   **1.** ans. will vary –4,6,10   **2.** ans. will vary – 1, 3, 5   **3.** 1536   **4.** 152   **5.** 18

Wed.   **1.** ans. will vary   **2.** ans. will vary   **3.** 1   **4.** 25 degrees, scalene   **5.** see reflection

Thurs.   **1.** 6   **2.** 63 min or 1h 3 min   **3.** =   **4.** 1 min   **5.** 5.4 cm

Fri.   **1.** Friday   **2.** Monday   **3.** 30   **4.** 20

Brain Stretch   sphere, cone, cylinder

Week 18

Mon.   **1.** ans. will vary   **2.** ans. will vary   **3.** no   **4.** 104, 94, 84   **5.** start at 144, subtract 10

Tues.   **1.** 6   **2.** 6.29   **3.** 45   **4.** 70   **5.** 48

Wed.   **1.** 6   **2.** similar   **3.** 5   **4.** square based pyramid   **5.** two shapes the same size and shape

Thurs.   **1.** scale   **2.** 21.6 cm2   **3.** a   **4.** 200 min   **5.** 104

Fri.   **1.** basketball   **2.** football   **3.** 22

Brain Stretch (based on 30 day month) **1.** 26 h   **2.** 338 h

Week 19

Mon.   **1.** 15   **2.** 35   **3.** 32   **4.** 17, 20, 23, 26, 29   **5.** 9x9 = 81

Tues.   **1.** 2360   **2.** 7   **3.** 2.618   **4.** 44   **5.** $62.49

Wed.   **1.** acute   **2.** right   **3.** see picture   **4.** ans. will vary   **5.** 12

Thurs.   **1.** 28   **2.** a   **3.** 6   **4.** 1400 m   **5.** measure with a ruler for accuracy

Fri.   **1.** 9   **2.** 12   **3.** 20 wpm   **4.** 11 wpm   **5.** 18

Brain Stretch   **1.** $9.58   **2.** $3.19, $3.19 , $3.20

Week 20

Mon.   **1.** no   **2.** 245, 250, 255   **3.** 93, 89, 85   **4.** 19   **5.** start at 1000, subtract 100

Tues.   **1.** 12   **2.** 8   **3.** 24   **4.** 24   **5.** ans. will vary

Wed.   **1.** ans. will vary   **2.** 6   **3.** straight angle   **4.** 2   **5.** 15 degrees, scalene

Thurs.   **1.** >   **2.** 96 units   **3.** 9:45 p.m.   **4.** 11:00   **5.** 1 h 20 min

Fri.   **1.** joke books   **2.** science fiction   **3.** joke, fiction, comic   **4.** 90   **5.** 21

Brain Stretch   8

Week 21

Mon. **1.** 147 **2.** 115 **3.** rectangle **4.** 31, 38, 45 **5.** growing

Tues. **1.** 4 **2.** 9000 **3.** 6/20 **4.** > **5.** $4.00

Wed. **1.** no **2.** ans will vary **3.** 2 **4.** none **5.** a

Thurs. **1.** 22 **2.** measuring cup **3.** 10 000 g **4.** 2.1 **5.** >

Fri. **1.** chocolate **2.** maple walnut **3.** vanilla, bubblegum **4.** 56 **5.** 14

Brain Stretch $66.60

Week 22

Mon. **1.** 8x4 – ans. may vary **2.** no **3.** 9 **4.** start at 900, subtract 50 **5.** growing

Tues. **1.** 7 R1 **2.** 11 R2 **3.** 40 **4.** 54 **5.** 5 groups R2

Wed. **1.** pentagon **2.** square **3.** e **4.** no **5.** cylinder – 2, 0, 0

Thurs. **1.** L **2.** a **3.** 28 min **4.** 50 **5.** 5:14 pm

Fri. **1.** 3/20 **2.** 2/20 or 1/10 **3.** 5/20 or 1/4 **4.** 12/20 or 3/5 **5.** 0/20

Brain Stretch   3

Week 23

Mon. **1.** 12 **2.** 79, 85, 91 **3.** true **4.** 202, 208, 214, 220 **5.** 48

Tues. **1.** 53 **2.** 31 **3.** 11/30 **4.** 27 **5.** 10

Wed. **1.** obtuse **2.** 2 **3.** **4.** 0 **5.** b

Thurs. **1.** 3.305 **2.** 1x18, 2x9, or 3x6 **3.** 24 **4.** a **5.** 3 min 20 s

Fri. **1.** d **2.** b **3.** a and c

Brain Stretch $12.00

Week 24

Mon. **1.** 90, 85, 79 **2.** → **3.** yes **4.** start at 99, subtract 6 **5.** 2

Tues. **1.** $29.97 **2.** 7397 **3.** 67 272 **4.** 63 **5.** 5.8

Wed. **1.** **2.** no **3.** congruent **4.** 1- if it's a right angle triangle **5.** slide or flip

Thurs. **1.** 21 **2.** 11 m **3.** metre stick **4.** 16, 11 **5.** 30 000

Fri. **1.** 23 **2.** 11 **3.** 18 **4.** 28 **5.** 5

Brain Stretch   **1.** 72 cards **2.** $6.24

Week 25

Mon. **1.** no **2.** 7 **3.** 37, 45, 53 **4.** 9 **5.** 60, 53, 46, 39, 32

Tues. **1.** 2469 **2.** forty-three dollars and eighty-nine cents **3.** 67/100 **4.** 10.1 **5.** 69

Wed. **1.** ans. will vary **2.** 4 **3.** i.e **4.** 10 **5.** c

Thurs. **1.** 730 **2.** 13 units^2 **3.** measure for accuracy **4.** > **5.** all sides will be 6 units

Fri. **1.** 226 **2.** 100 m sprint **3.** high jump **4.** 91 **5.** 46

Brain Stretch   3.9 min

Week 26

Mon.   **1.** start at 60, add 15 **2.** no **3.** c **4.** ans. will vary **5.** 55

Tues.   **1.** 8 **2.** 3 x toonie, 1 x loonie, 2 x quarter, 2 x dime, 1 x penny **3.** .89 **4.** 390 **5.** $7.12

Wed.   **1.** 0 **2.** drawing: an angle less than 90° **3.** ←——⌣——→ **4.** 3 possible, drawn through each vertex

        5. circle – no angles, no sides, circular lines, no vertices **triangle** – 3 angles, 3 sides, 3 vertices

Thurs. **1.** > **2.** a **3.** February 20th **4.** 600 **5.** 2.680

Fri.   **1.** week 1 **2.** $22.00 **3.** $78.00 **4.** $6.00

Brain Stretch   **1.** 720 m² **2.** 108 m **3.** $864.00

Week 27

Mon.   **1.** 220, 209, 198, 187, 176, 165 **2.** 10 **3.** b **4.** start at 2, add 11 **5.** 88, 87, 89

Tues.   **1.** 9801, 8901, 8109 **2.** > **3.** 2000 **4.** 2/3 **5.** 7/8

Wed.   **1.** 1 **2.** obtuse **3.** 0 **4.** ⇒ **5.** triangles

Thurs. **1.** 49 dm² **2.** 208 **3.** 6 minutes **4.** ruler **5.** each side is 5 units

Fri.   symbol will vary, key = 2

Brain Stretch   80

Week 28

Mon.   **1.** 4 **2.** 59, 57, 55 **3.** no **4.** 56 **5.** b

Tues.   **1.** 3/6 **2.** 523 **3.** two thousand seven hundred forty-five **4.** 48 **5.** 7

Wed.   **1.** 4 **2.** see reflection **3.** 0 **4.** 0 **5.** ans. will vary

Thurs. **1.** a **2.** horse **3.** 30 m **4.** 41 **5.** 2190

Fri.   **1.** 100 **2.** 20 **3.** 100 **4.** 20 **5.** 80

Brain Stretch   **1.** 80 **2.** about 4

Week 29

Mon.   **1.** ans. will vary **2.** 200, 185, 170 **3.** a **4.** 15 **5.** 8

Tues. **1.** 0.1289 **2.** 6/10 **3.** 334, 345, 435, 545 **4.** $5.00 **5.** 1, 2, 3, 4, 6, 8, 9, 12, 18, 24, 36, 72

Wed.   **1.** false **2.** ans. will vary **3.** ▱ **4.** yes **5.** ans. will vary

Thurs. **1.** 24 mm **2.** 4 **3.** ans. will vary – 9 x 8 **4.** check for accuracy with a ruler **5.** 12 h

Fri.   **1.** 600 **2.** comedy **3.** 600 **4.** action and horror **5.** 3450

Brain Stretch   4000 m

Week 30

Mon.   **1.** 1728 **2.** no **3.** 16 **4.** start at 32, add 4 **5.** 135 km

Tues. **1.** 1, 2, 3, 4, 6, 8, 12,16, 24, 48 **2.** 6 R2 **3.** 60 000+7000+500+20+3 **4.** 48342 **5.** 9900

Wed. **1.** no **2.** true **3.** 12 **4.** ⇦ **5.** 2 figures that have the same size and shape

Thurs. **1.** .11 m **2.** < **3.** 560 min **4.** 8 m² **5.** clock

Fri.   **1.** spinner 1 **2.** spinner 2 **3.** spinner 1 **4.** d

Brain Stretch   **1.** 15 **2.** $45.00 **3.** 45